Josimar Vieira dos Reis
Elisabeth Regina Alves Cavalcanti Silva
José Gustavo da Silva Melo

Payment for Environmental Services

Josimar Vieira dos Reis
Elisabeth Regina Alves Cavalcanti Silva
José Gustavo da Silva Melo

Payment for Environmental Services

A look at the incentive instrument for environmental conservation

ScienciaScripts

Imprint

Any brand names and product names mentioned in this book are subject to trademark, brand or patent protection and are trademarks or registered trademarks of their respective holders. The use of brand names, product names, common names, trade names, product descriptions etc. even without a particular marking in this work is in no way to be construed to mean that such names may be regarded as unrestricted in respect of trademark and brand protection legislation and could thus be used by anyone.

Cover image: www.ingimage.com

This book is a translation from the original published under ISBN 978-613-9-64308-0.

Publisher:
Sciencia Scripts
is a trademark of
Dodo Books Indian Ocean Ltd. and OmniScriptum S.R.L publishing group

120 High Road, East Finchley, London, N2 9ED, United Kingdom
Str. Armeneasca 28/1, office 1, Chisinau MD-2012, Republic of Moldova, Europe
Printed at: see last page
ISBN: 978-620-7-76518-8

Copyright © Josimar Vieira dos Reis, Elisabeth Regina Alves Cavalcanti Silva, José Gustavo da Silva Melo
Copyright © 2024 Dodo Books Indian Ocean Ltd. and OmniScriptum S.R.L publishing group

I dedicate this research to God, my mother and my family and friends for their support during its preparation.

ACKNOWLEDGMENTS

To **God**, for life, protection and the strength to overcome obstacles.

To my **mother**, for everything she taught me.

To my brother **Ygor Rafael** (in memoriam).

To all the teachers on the course for their learning throughout the subjects.

To my course friends for their companionship throughout the course.

To the Course Coordinator **Ana Maria Ataides**. For her affection, dedication and companionship.

To my advisor Prof. MSc. **Midia da Silva Rodrigues** for her guidance and companionship.

To all those who contributed directly and indirectly to this work.

Veni, vidi, vici ["I came, I saw and I won"].
Emperor Julius Cdsar

SUMMARY

This paper is based on the main aspects of payment for environmental services (PES). Payment for Environmental Services (PES) is considered a promising mechanism for solving some problems related to the degradation of ecosystems by anthropogenic action. PES occurs when beneficiaries are willing to pay providers who adopt conservation practices or preserve forested areas. In this paper we will look at its main aspects and definitions, such as: how to pay, its challenges and its implementation as an economic instrument, its effectiveness and efficiency in the face of the economic market, its social aspects and its value for protecting ecosystems. The research is a systematic review of the literature, and sought more general information on the subject of study, to later involve a study of the foundations and main aspects of the mechanisms of payment for environmental services. The conclusion is that this is a promising project, with the potential to be equitable, effective and efficient. This already represents recognition of its importance, as well as a step forward for environmental conservation. In the same way, the links inferred can support the structuring and public management of future public policies and PES projects.

Keywords: payments for environmental services (PES), economic instruments, public policies.

SUMMARY

1 INTRODUCTION

Environmental problems have been the general concern of humanity and preserving the environment is today a question of survival, of guaranteeing survival for present generations and making life possible for future generations.

According to Spironello (2008), the search for solutions to environmental problems and the beginning of society's awareness of the need to rationalize the use of natural resources began in the 1960s. Parallel to this process is the idea of land-use planning, whose main objective is to improve people's quality of life.

For some decades now, environmental issues have become a recurring theme in society, since a planetary environmental crisis is increasingly imminent according to the reports and technical opinions developed by experts around the world. Since the 1970s, with the Stockholm Conference, considered a milestone in terms of environmental reflection, the understanding of environmental awareness has spread significantly to the present day.

In short, everyone is in favor of preserving nature, but few actually do anything about it. There are serious threats to life on the planet, given the high level of environmental degradation we have reached in this risk society (BECK, 2010, p. 12). In the dead of night, when everyone is asleep, forests are devastated, rivers, streams and lakes are polluted, wild and domestic animals are killed and, the next day, everything is justified in the name of the need to occupy spaces, seeking to build industries, homes, bridges, roads, in short, everything in the name of progress.
According to Engel, Pagiola and Wunder (2008), in general, ecosystem services can be defined as the benefits that individuals obtain from natural ecosystems. Between 1960 and 2000, the demand for ecosystem services increased significantly, as the world's population doubled and the global economy grew more than sixfold.

Faced with the high level of degradation of nature, the challenges for environmental preservation are diverse and involve complex issues such as excessive pollution, climate change, global warming, water pollution, solid waste, sustainable development, basic sanitation, environmental refugees, nuclear energy, chemical/toxic waste, extinction of species and biodiversity, scarcity of fresh water, etc.

We can only act irresponsibly when we take responsibility [...] the use we make of natural resources, water, soil, vegetation, we are responsible for their conservation, but when we use them inappropriately, we are being irresponsible with what nature offers us for our present and future existence on earth (SANTOS; GUIMARAES, 2010).

There are assets, such as riparian forests, whose need for preservation is indisputable, as they mean a guarantee of life, biodiversity, the quality of drinking water and the survival of human beings, fauna and flora. According to the Constitution of the Federative Republic of Brazil:

Art. 225: Everyone has the right to an ecologically balanced environment, which is a common good and essential to a healthy quality of life, and the public authorities and the community have a duty to defend and preserve it for present and future generations (BRASIL, 1988).

This constitutional rule, therefore, enshrines the preservation of the environment by assigning responsibilities to the public spheres to ensure environmental rights; decentralizing competences, understanding that it is municipalizing with a view to ecological protection.

The aim of this work is to develop the theme of preserving and restoring the environment through financial incentives for environmental preservation, combining active conduct by the state and/or society, embodied in a Payment for Environmental Service - PSA (financial counterpart).

According to Gusmao (2008), payment for environmental services is a strong instrument for achieving sustainable development, extending to the global environment, culminating in the inclusion and improvement of the well-being of forest producers.

An extremely topical and exciting issue in the debate on sustainable development is the PES, an instrument that pays or rewards producers who protect forests. It is a way of stimulating conservation, attributed to the sustainable exploitation of forests, contributing to the social reproduction of traditional populations, through a more equitable distribution of income.

According to Rojas and Aylward (2005), ecosystem services refer to natural systems that provide a continuous flow of goods and services to society. Environmental services can include hydrological benefits, sedimentation reduction, disaster prevention, biodiversity conservation and carbon sequestration.

According to Pagiola, Arcenas and Plantais (2005), PES is an instrument suggested by the United Nations, already successfully adopted in several countries, and has been the subject of programs being implemented in some Brazilian municipalities, suggesting complementing the command and control actions so far adopted by the Brazilian state with a reward instrument, This payment, in addition to environmental preservation and recovery, stimulates cooperation, promotes a better distribution of the cost of conserving ecosystems, recognizes and rewards, stimulating conservationist conduct and, consequently, the self-esteem and dignity of the preserver. It helps keep people in the countryside, values their actions and, finally, promotes better integration and efficiency in environmental conservation actions.

According to Wundere Kanounnikoff (2009), with its emergence, several definitions and questions about the instrument arose at the same time, reinforcing the need for this mechanism. Despite the numerous definitions of PES, one of them is widely used and defines it as "a voluntary transaction in which a well-defined environmental service, or a land use that can ensure this service, is acquired by at least one buyer from at least one provider, under the condition that the provider guarantees the provision of the service (conditionality)".

The financial incentives provided to providers of environmental services are mainly related to the control of environmental problems. This is because forest cover protects the banks of water bodies, reducing environmentally damaging processes such as siltation of rivers, thus playing a direct role in improving the quantity and quality of water; as well as preserving fauna and flora, among other subsystems that involve the environment.

According to the Food and Agriculture Organization - FAO (2004), Payments for Environmental Services schemes are flexible compensation mechanisms whereby the providers of environmental services are paid by the users of these services. PES are considered promising mechanisms for financing environmental protection and restoration, as well as a way of complementing and reinforcing existing regulations.

Most existing PES schemes work with four main groups of environmental services: scenic beauty, carbon sequestration, biodiversity conservation and watershed protection (LANDELL-MILLS; PORRAS, 2002).

In this way, implementing a program to remunerate the environmental services provided by those who safeguard the increasingly restricted environmental factors essential to the balance of the natural environment is a valid strategy that can provide an "escape valve" for human intervention in the natural environment, and can help in discussions about effective sustainable development.

2 OBJECTIVE

The aim of this work is to carry out a bibliographical survey on the foundations and main aspects of payment for environmental services, covering its definitions, forms of payment, its challenges and its implementation as an economic instrument, its effectiveness and efficiency in relation to the economic market, its social aspects and its value for the conservation of ecosystems.

3 METHODOLOGY

For the development of this monograph, a literature review was carried out, which included materials published from 1960 to 2014. The information was obtained through a literary survey of books, periodicals, websites, articles and specialized magazines, various documents such as reports, plans and legislation related to the proposed topic.

The purpose of bibliographic research is to find out about the different forms of scientific contribution that have been made on a given subject or phenomenon. It is based on material that has already been published, consisting mainly of articles from scientific journals, books, course completion papers and material available on the Internet (VASCONCELOS, 2005).

The search strategy for articles included a search in electronic databases and a manual search for citations in the publications initially identified. The following electronic databases were used: Scientific Electronic Library Online (SCIELO), Digital Library of the Environment - IBAMA, Ministry of the Environment - MMA, Secretariat of the Environment of Pernambuco - Semas.

The manual search was carried out in the libraries of the Federal University of Pernambuco; Federal Rural University of Pernambuco; Federal Institute of Pernambuco; Frassinetti College of Recife and, in addition, direct contact with some authors via e-mail made it possible to obtain several articles.

4 CHAPTER 1

4.1 The historical evolution of the use of public policy instruments for environmental conservation

4.2 World Conferences

The global context of concern for future generations regarding natural resources intersects with international law and the environment, and is marked by the 1972 United Nations Conference on the Environment (Stockholm), with the Stockholm Declaration, and the Brundtland Report (1987). In this sense: "Sustainable development is development that meets present needs without compromising the ability of future generations to meet their own needs" (WCED, 1987).

For Romeiro (2001, p.8) the expression "Development

Sustainable" emerges from this context as a

a conciliatory proposal in which it is recognized that technical progress effectively relativizes environmental limits, but does not eliminate them, and in which economic growth is a necessary but not sufficient condition for the elimination of poverty and social disparities.

It was through these conventions that the Convention on Climate Change and Biological Diversity and an action plan called Agenda 21 emerged, which were not intended to be binding, but rather guidelines. Through these standards, important principles such as international cooperation, precaution and prevention were defined.

In this sense, the environmental issue at the international level also has some other milestones: in 1992 in Rio de Janeiro (Rio 92) with the Declaration on Environment and Development; the Convention on Biological Diversity (CBD), the Kyoto Protocol (1997) and the United Nations Framework Convention on Climate Change. Also noteworthy is the so-called REDD - Reduce Emissions for Deforestation and Degradation.

The latter aims to reduce carbon emissions, greenhouse gas emissions and those that are avoided by reducing deforestation and forest degradation, especially in tropical forests. It originated through a proposal known as "Compensated Emission

Reduction" (SANTILLI et al, 2000), at COP-9 in 2003 in Milan, Italy (2003), in which researchers from the Amazon Environmental Research Institute (IPAM, 2013) took part.

REDD+ today refers to a mechanism or policy that can provide positive incentives for developing countries to reduce emissions. They therefore include actions to mitigate climate change, namely: a) reducing emissions from deforestation and forest degradation; b. increasing forest carbon reserves; c. sustainable forest management; d. forest conservation (IPAM, 2013).

According to Leonardo (2003), one of the main reasons for the degradation of natural resources that are fundamental to human survival, such as soil and water, has been the irrational use and management of the land, without prior assessment of its potential and limitations.

We need to redefine the pace, form of consumption and exploitation of natural resources, man's relationship with nature, as well as population growth in order to guarantee living conditions not only for humanity, but also for other species.

For Rezek (2008), development should not be sought at the cost of environmental sacrifice, because it will not last; but it is unfair and biased to claim that environmental preservation should act as an obstacle to the development of poor nations or those that have not yet fully achieved it. Reconciling the two values is the concept of sustainable development as that which does not sacrifice its own scenario, that which does not compromise its own conditions of durability.

Faced with the growing pressure on ecosystems, various institutions and governments have sought to create incentives to improve the management of environmental assets. In this sense, Payment for Environmental Services policies have been identified around the world as a viable option for achieving this goal. These policies can complement command and control instruments, helping to increase the value of environmental assets, as well as bringing benefits to the providers of these services (especially vulnerable populations), improving their quality of life (FOREST TRENDS; KATOOMBA GROUP, 2008).

4.3 Theoretical Basis

The theoretical basis of payment for environmental services (PES) schemes is not new, with the key concepts of externalities and public goods dating back to at least the beginning of the 20th century. However, it is only in the last few decades that PES has been gaining ground in publications around the world, as well as serving as the basis for various practical public policy experiments.

PES schemes are derived from the Coase Theorem of 1960, which states that through negotiations agents internalize externalities and achieve efficiency, regardless of the initial allocation of property rights and in the absence of transaction costs (KOSOY et al, 2007).

The Coase Theorem (which is actually a view/constraint and not a theorem) was formulated by Ronald Coase (American economist) and states that externalities or economic inefficiencies can, under certain circumstances, be corrected and internalized by negotiation between the affected parties, without the need for intervention by a regulatory body. The circumstances required for this to be possible are, according to Coase (1960), the possibility of negotiation without transaction costs and the existence of guaranteed and well-defined property rights.

According to Coase (1960), externalities occur when a person acts causing effects on other people, without their consent, and the effect can be beneficial - positive externality - or harmful - negative externality.

For Hercowitz and Whately (2008), externalities are secondary effects generated as a result of a decision made by an individual economic agent which affect other economic agents who do not take part in the decision. They can be positive or negative.

So, for example, suppose that an industry uses water in its production process to cool its machinery and returns the water to the river in a different state from the one it was taken from, with poorer quality. Suppose, furthermore, that this water returned to the river in poorer quality will affect a farming community downstream, which will see its production compromised. In this case, the industry is causing a negative externality that is perceived by the farming community.

4.4 PSA Worldwide

Currently, the discussion on Payment for Environmental Services is still being shaped at the major conferences on climate change and biodiversity, but some PES models have already been implemented in some parts of the world.

Few "mature" PES programs exist in Asia, but interest in the subject is gradually growing. The largest number of PES or PES studies on the continent come from Indonesia and the Philippines, where watershed management has replaced the command and control approach (TRES, 2011).

The total number of PES in Latin America exceeds the numbers in Africa and Asia. This is due to more secure rural land tenure in that region, as well as greater acceptance of land use trading rights and land management practices (TRES, 2011). In Latin America, there is a target of implementing 32 funds by 2015.

Most of the countries with watershed-related PES are developing countries. Of the developed countries, the only ones mentioned are the United States and France, both of which are benchmarks when it comes to environmental policies.

Wunder, Engel and Pagiola (2008) point to a difference in the way programs are funded between developed and developing countries. Developed countries often receive funding from various levels of government, while developing countries generally receive grants. They point out that in developing countries there is a great contrast between public and private programs, especially in terms of the differentiation of payments.

Kawaichi (2009) surveyed the use of environmental policies adopted by various countries and observed that the greatest frequency of use of the PES mechanism occurs in developing countries, although the use of PES is facilitated in developed countries due to the greater availability of data.

Although there are more projects related to water, the ones that are already in the implementation phase are more related to carbon sequestration (15 carbon, 8 water and only 1 biodiversity).

4.5 Payment for Environmental Services in Brazil - Some Examples

Brazil, due to its territorial extension and the presence of large forest remnants, as well as a growing increase in environmental awareness, has fertile ground for the application of PES. In this section, some Brazilian initiatives will be discussed in order to exemplify and analyze the programs implemented in the country in order to provide subsidies for the discussion of the subject and even the implementation of future projects.

For Wunder, Engel and Pagiola (2008) in Brazil, PES has been discussed with more attention since the launch of the Proambiente Program in 2000, which was an initial experience of PES in the country, but demonstrated several challenges to be overcome.

Since then, several bills have been proposed in Congress on the subject and some federal laws already mention it, although they do not create a national regime in this regard. In addition, some states have published laws dealing with the issue, but there has not yet been a comparative analysis of how these already approved laws address the issue.

Policy instruments are mechanisms used to achieve public policy objectives. According to Mota (2006), public policies consist of government actions to intervene in the economy, without which economic agents alone would not be able to intervene. Public policies are therefore means of correcting market failures.

In the mid-1990s, market-based mechanisms emerged as a response to the rapid loss of vegetation cover and consequent loss of the services provided by forests, exemplified by water flow regulation, carbon storage and biodiversity.

Among the examples to be cited are PROAMBIENTE, the Ecological ICMS, the Water Producer Program, an initiative of the National Water Agency, the action of The Nature Conservancy (TNC), the Guandu Hydrographic Basin Committee (RJ) and the Extrema Water Conservancy Project (MG).

In the state of Pernambuco, the Agua do Parque project, carried out by the Northeast Environmental Research Center (CEPAN) in partnership with the Federal Institute of Education, Science and Technology of Pernambuco (IFPE) and the Federal University of Pernambuco (UFPE), is coming to an end with good results. Through studies carried out as part of the project, it was possible to identify that Companhia Pernambucana de Saneamento (Compesa) saves between R$9,000 and R$11,000 on the treatment of water extracted from the Dois Irmaos State Park.

Agua do Parque, which received funding from the Brazilian Biodiversity Fund (Funbio), installed equipment and carried out measurements in the Dois Irmaos State Park, with the aim of evaluating the environmental services provided by the park and the influence of the preserved Atlantic Forest on the quality of the local water. The studies evaluated treatment costs at four Water Treatment Plants (WTPs), with an emphasis on evaluating costs related to turbidity, i.e. the ability of light to penetrate the water. The cleaner the water, the lower the turbidity and the greater the light penetration. The study showed that, as a result of the lower turbidity, Compesa saves R$0.03 (three cents) for each cubic meter of water treated, which means that the water in Parque Dois Irmaos is clean and of good quality. This quality indicates that the forest is generating a benefit for society as a whole by maintaining the purity of the water and generating savings.

The organization intends to replicate the experience of the Water in the Park project in other conservation units in the state and in the Northeast Biodiversity Corridor. In the long term, the NGO and its partners intend to carry out broader studies to demonstrate other benefits such as the beauty of the landscape, temperature variation and carbon capture that the Conservation Units provide to society.

The law is considered to be a pioneering move by the state and will bring benefits such as the possibility of quantifying the value of natural capital, as well as the state's environmental liabilities; identifying the beneficiaries and providers of environmental services; sharing responsibility for conservation; and generating opportunities for new green markets.

Authors Pagiola, Landell-Mills and Bishop see market-based instruments as efficient incentives to conserve forests, and their results can be verified more quickly when compared to command and control policies.

The most frequent environmental services in these programs are those related especially to water, such as regulating the flow of water sources and maintaining water cycles. In fact, the main PES program underway in the country, the Water Producer Program, is focused on water protection in Brazil.

According to Almeida (1998), "the main characteristic of the command and control policy" is that, on a legal basis, it treats the polluter as an "eco-offender" and, as such, does not give him a choice: he has to obey the imposed rule, otherwise he is subject to penalties in judicial or administrative proceedings.

The imposition of fines in cases of non-compliance with the obligation is quite common. It can be seen from the above that command and control instruments induce behavior deemed optimal according to the State's determination, and changes in the behavior of polluting agents are induced through the imposition of obligations, which, if not complied with, will be penalized through judicial or administrative proceedings. According to Varela (2001, p.13), these instruments:

The aim of these mechanisms is to reduce regulation, give the agents involved greater flexibility when it comes to alternatives, reduce the costs of controlling environmental problems and stimulate the development of cleaner technologies. They can be called 'polluter pays' mechanisms, when the instrument used makes the polluter pay for the damage caused, or 'user pays', when it is the user who has to pay for the total social cost that the product generates for the environment.

There is the possibility of achieving the proposed objectives at a lower cost through regulatory measures, while at the same time creating incentives for innovation and improvement. The main reason for this lies in the different costs of the situations encountered, leading economic agents to always prioritize those situations in which the cost would be lower and the return higher.

In Brazil there is a bill that establishes the National Policy for Environmental Services, the Federal Program for Payment for Environmental Services, establishes forms of control and financing of this Program, known asPL 792/2007.The proposal, presented by deputy Anselmo de Jesus (RO), defines the concepts, objectives and guidelines of the National Policy for Payment for Environmental Services, in addition to creating the National Commission for the Policy for Payment for Environmental Services; the Federal Program for Payment for Environmental Services; and the Federal Fund for Payment for Environmental Services.

The bill defines environmental services that can be paid for as individual or collective initiatives that can favor the maintenance, recovery or improvement of environmental or ecosystem services. Among other things, it establishes priority for payment for environmental services provided in ecosystems at greater socio-environmental risk.

For Bourdieu (2004), however, public policies can only produce the desired effect if they are internalized by individuals and become a habit. This, in the sociologist's words, means "the incorporated, almost postural disposition of acquired knowledge", i.e. a socialized body. In other words, public environmental policies must take on a pedagogical stance in the sense of being incorporated - internalized - by individuals into their social practice.

5 CHAPTER 2

5.1 Main ecosystem services

According to Nusdeo (2012), environmental services are varied and the subject of this study is those that support living conditions, which are also many and diverse. Examples of the latter include natural pollination, soil nutrient cycling, maintaining the volume and quality of water resources and carbon sequestration, which stabilizes the climate.

In order to discuss the possibility of remunerating the service, however, it is necessary to mobilize a group of agents willing to pay for the service from identified groups of providers. This mobilization involves factors such as increased scientific certainty about the service, the perception of key players in the process about its importance and even the interest of some groups in developing a market for such services, which will lead to the structuring of this market and the corresponding payments.

Carbon credits are a good example of the simultaneous occurrence of these elements, with a major boost given by intermediaries to the relationship between buyers and service providers. For this reason, the experiences of remuneration for environmental services currently take place predominantly around a group of four environmental services: biodiversity conservation, watershed protection, carbon sequestration and storage and scenic beauty, sometimes with the possibility of providing some of them together.

It is difficult to observe an environmental service being provided in isolation, which means that conserved areas, for example, provide numerous interconnected services, such as: biodiversity conservation, contributing to improving the water body, maintaining nutrient cycling, regulating the microclimate, among others (Institute for Applied Economic Research - IPEA, 2010). This is perhaps one of the limitations of PES, as this instrument requires a well-defined environmental service to be provided. However, it is not always possible to clearly delimit the service.

Wund and Wertz-Kanounnikoff (2009) cite four main ecosystem services used in payments for environmental services: i) Carbon sequestration, ii) Biodiversity protection, iii) Watershed protection, iv) Landscape beauty.

It does not ignore the fact that there are many other services to be considered, but only these four are of significant commercial scale. This includes a wide range of ecosystems, from relatively untouched ones, such as natural forests, through landscapes with different forms of human use, to intensively managed and modified ecosystems, such as agricultural and urban areas. May and Geluda (2005) complement this concept by referring to these as benefits generated for society by nature, which until then had not been monetarily remunerated to their providers.

5.2 Characteristics of Environmental Services

Environmental services have the characteristics of public goods: non-exclusivity and non-rivalry. Because of this, property rights are not well defined (SEROA DA MOTTA,1998).

The fact that these two characteristics of public goods make it impossible to set prices creates a market failure that prevents the efficient allocation of resources. As there are no nails, there is no reference to the scarcity of resources, leading to the tragedy of the commons. As a result, the problem of the hitchhiker arises and there is an overexploitation and underproduction of services. It turns out that those who "produce" them don't get paid for them and those who "consume" them don't pay for them.

Guedes and Seehusen (2011) point out that not all environmental goods and services suffer from the characteristics of pure public goods, there are different levels of non-exclusivity and non-rivalry, and this intensity will determine the degree of market failure.

For Wunder (2007) it is very complex to determine the precise relationship between land uses and environmental services and there is little knowledge regarding the implementation of PES systems.

Defining the product to be marketed is still one of the most challenging aspects in the field of environmental services (LANDELL-MILLS; PORRAS, 2002).

The management options to get around this market failure are mainly focused on command and control instruments and economic incentives. The former are regulatory, requiring economic agents to achieve the imposed targets regardless of the costs. Economic incentives aim to ensure that economic agents incorporate the costs or benefits generated into their decisions. To internalize externalities, which is the premise of economic incentives, there are two alternatives: the Pigouvian and the Coasean. The Pigouvian approach suggests imposing taxes or subsidies to compensate for environmental costs or benefits. The idea behind the tax is to correct the market value in such a way that it comes to represent the social value.

It is complicated to implement because of the difficulty in calculating the value of the environmental damage in full.

According to Seroa da Motta (1998), this negotiation makes it possible to achieve environmental objectives at lower costs and maximizes the aggregate social gains.In PES systems, the idea of Coase has been adopted (GUEDES; SEEHUSEN, 2011).

5.3 Payment methods

According to Sommerville, Jones and Milnergullano (2009), PES compensation is intended to be a transfer of positive incentives, financial or otherwise, whose impact provides gains to the providers of environmental services. Muradian et al. (2010) say that cash payments have limited impacts and that, if they are too small, they can discourage providers and diminish ethical incentives to conserve. If they are prolonged, they lose their incentive character and become seen as an entitlement. Some say that monetary rewards can end up disrupting a pre-existing social market based on social lakes and reciprocity. In undesirable situations, the effects on conservation are better when there is no payment than when there are derisory payments.

Mota (2006) states that all goods have economic value because they have a price set by the market. But biodiversity resources, such as an orangutan, a forest, the air and many others, have no price set by the markets. These natural resources are not commodities; they are essential assets for the preservation of life for all beings.

The sustainability of the PES contract often depends on its effects on community income, changes in consumption and demand for land and labor. Many development professionals are hesitant to defend the direct transfer of money to rural communities, as they doubt the ability of money to generate sustained social welfare. They claim that money can increase alcohol consumption, unnecessary goods and cause social disorders such as stress.

However, the regular transfer of money is more effective in reducing poverty than other forms of payment. There are many ways of compensating for the provision of services other than money, such as the provision of public goods for the community, the implementation of infrastructure in the region, technical assistance, equipment, training, among others.

Heyman and Ariely (2004) say that low-value inkind payments are more effective than low-value in-kind payments, since recipients assimilate this type of payment as an exchange, an act of reciprocity, as if they belonged to social markets.

When the payment is of the inkind type, there is a greater chance of structural changes occurring in the local economy. The providers of environmental services can be trained to carry out other, more "refined" activities, and there can also be a change in the pattern of use of natural resources (IPEA, 2010).

Costa (2008) observed that compensating farmers for the opportunity costs of avoided deforestation from a social and economic point of view may be less promising than providing conditions for family farmers to switch to alternative land uses that provide lower levels of environmental services compared to avoided deforestation, but are more beneficial in economic and social terms, promoting the development of family farming in a more sustainable way.

Even considering that environmental services and, consequently, the attribution of values to them, only make sense when it comes to their relationship with man, it is important to emphasize Ortiz's (2003) assertion that every environmental resource has an intrinsic value which, by definition, is the value that is proper, internal, inherent or peculiar to it. It is the value that reflects the existence rights of non-human species and inanimate objects, for example.

For Veiga Neto (2008), the discussion about paying for the services provided by ecosystems began with some important points. The first relates to society's growing perception of the constant deterioration of these services, which concluded that more than 60% of the world's ecosystems have been used in a way that is not beneficial to the environment. Payments for Environmental Services are financial transfers from the beneficiaries of environmental services to those who, because of practices that conserve nature, provide these services.

PES can promote conservation through financial incentives for the providers of environmental services. This model complements the user-pays principle by focusing on the provision of the service: it is the "provider-recipient" principle, in which users pay and conservationists receive (ANA, 2012).

Engel, Pagiola and Wunder (2008) warn of the effects that PES programs can have on local economies. They mention the importance of having a prior study characterizing the changes that may occur after the implementation of the PES in the community's income. This is because the type of activity determined by the program has a series of impacts, not only on land cover and the costs of maintaining the program, but also on local economies, since they influence the demand for labor and the generation of income. For example, where there are payments to keep a forest standing instead of turning it into arable land, this means a loss of demand for labor and a reduction in income. However, when it comes to areas degraded by extensive agriculture and there is payment to transform that site into an area of intensive forestry and pastoral practice, there is an increase in demand for labor that strengthens the local economy (ENGEL; PAGIOLA; WUNDER, 2008). This interference in the economy can represent a difficulty in implementing PES programs, since the program may not be able to be implemented.

accepted by the community precisely because it diminishes the local labor market and limits the region's economic development.

5.4 PES as an Economic Tool

According to Loureiro (2008), the growing use of economic instruments (EIs) for biodiversity conservation, in addition to traditional command and control instruments, is a worldwide trend.

PES is a market-based economic instrument, i.e. a market mechanism (IPEA). This means that it must allocate resources efficiently; however, this does not necessarily entail a sustainable condition, nor does it justify the distribution of resources among agents. For these two situations to occur, there needs to be government intervention in defining the scale to be adopted in the program, as well as the introduction of additional rules and instruments that take into account the desired distributive aspects (IPEA, 2010).

For Guedes and Seehusen (2011), the benefits generated by the environmental services provided have a main level of reach, and it is based on this level of reach that the markets for the services are formed. For example: water quantity and quality have their benefits established more at the local level, while sources of raw materials and food are regional and biodiversity conservation, global. This main level of scope will also help determine possible sources of funding.

The idea of Payment for Environmental Services stems, on the one hand, from the recognition that ecosystems effectively provide important services that must be conserved and, on the other hand, from the understanding that as long as these services are not part of the market, i.e. do not have a monetary value, they will not be part of the decision making of the agents who relate to these services and, consequently, run the risk of being extinguished in favor of other profitable activities.

For Guedes and Seehusen (2011), this conservation mechanism is expected to increase the relative profitability of sustainable use and protection activities compared to undesirable activities.

Wunder and Wertz-Kanounnikoff (2009) make a comparison between PES and other instruments that promote conservation. This comparison is important because it helps to decide which instruments are the most suitable for each situation, making it easier for them to operate efficiently.

According to Mota (2006), in addition to aiming to conserve natural resources, PES schemes can also maximize social welfare, since they are economic instruments and therefore aim for efficiency, financing a social activity in which the market price is corrected to finance a given level of revenue, in order to cover provision costs or investments in environmental protection services, and induce social behavior, with the intention of correcting the market price of a good or service, in order to induce a change in the behavior of the economic agent, towards a more efficient pattern of use of the resource, without having the main objective of generating revenue.

Furthermore, it is an innovative initiative that is attracting attention in both developed and developing countries. After all, linking Payments for Environmental Services to economic development and poverty reduction is an important issue in developing countries, as PES can represent a new source of financial support for their economic and environmental development through the use of funds from the global community. It is important to note that in addition to the impact of payments on employment and income, there can be significant economic development benefits associated with the environmental service itself.

Foleto and Leite (2011) say that the novelty of PES mechanisms is the fact that they do not use the government's imposing force to comply with legislation, but rather involvement, encouraging landowners and society to reflect on the importance of maintaining natural ecosystems, benefiting economically those who help to conserve them.

Although Payments for Environmental Services are potential drivers of economic development in the societies that apply them, they are not welfare instruments. It is clear that they consist of initiatives to remunerate private landowners who limit their production in exchange for conservation.

According to Engel, Pagiola and Wunder (2008), in general, economic incentives are more efficient than pure regulation or command and control. This is due to the fact that command and control instruments determine the same level of activity for all OS providers, while economic instruments are more flexible. For example, the desire to conserve forests would apply to all forests, regardless of their level of productivity or the cost of conservation.

An important point in PES projects is the definition of the prices to be paid for environmental services, especially in the case of water and biodiversity. As there are no established markets for these services, the value of the payments must be negotiated between the buyer and the provider of the environmental services in order to arrive at a fair and viable value.

Economic valuation is not strictly necessary to arrive at the amounts to be paid. However, it can be very useful in helping to determine them. On the willingness-to-pay side, it can show buyers an estimate of the economic benefits related to the provision of each environmental service. On the supply side, it can help estimate the additional costs incurred by producers, which depend on opportunity costs and the difference in cost between sustainable and less sustainable practices. These additional costs often determine the minimum willingness of producers to enter into a PES scheme.

6 CHAPTER 3

6.1 Situations suitable for PES application

According to Borner, Hohnwald and Vosti (2008), when small payments to landowners make a difference in the decision to adopt the desired land use, when scenarios present foreseen environmental problems, thus payments are a kind of environmental insurance, when the sum of the expected benefits for society is greater than the expected costs of the action, related to the implementation, maintenance, monitoring of the instruments adopted.

For Wunder, Engel and Pagiola (2008), some conditions are necessary for PES schemes to work, such as economic, cultural, institutional and informational preconditions. The economic precondition refers to the existence of an externality (an external benefit) that must be compensated for.

The concept of Opportunity Cost is particularly useful for evaluating alternatives when the goods involved are not tradable, such as education, health, safety or environmental services. In a production process for environmental goods or services, the opportunity cost of an environmental factor corresponds to the best gain that could be obtained by using that factor instead of another activity other than production. The environmental opportunity cost is the maximum value that could have been obtained by using an environmental resource.

For Seehusen (2007), this makes it possible for a market to emerge with demanders and providers of environmental services, which can lead to PES systems.

According to Engel, Pagiola and Wunder (2008), activities that lack additionality are not inefficient from a social point of view, since the desired practice has been adopted. The crux of the matter is economic inefficiency, since remuneration has been spent that would not have been necessary to achieve the objective. However, this inefficiency can lead to social inefficiency if there are limited funds for the program.

The money spent there could no longer be used in another situation that would not have been possible without the payment. It also wastes transaction costs, which would have been unnecessary without the PSA.

There are several factors that define whether a PES project will be successful or not. Firstly, it is necessary to investigate to what extent the PES instrument is more appropriate than others for achieving the desired environmental objectives.

It is advisable to use PES as a management option if the benefits generated by the instrument (the improvement in the provision of environmental services) are greater than its implementation costs (costs of managing the mechanism, field activities, awareness-raising and coordination, etc.). If this is not the case, it is advisable to analyse whether there are more cost-effective management options for dealing with the environmental problem, the transaction costs of which are minimized by government regulation due to the larger scale of operation. In the case of voluntary agreements, management costs are lower because the agreement is simple, without bureaucracy, and therefore less costly.

According to Engel, Pagiola and Wunder (2008), flexible payments allow more people to participate in the program with the same total expenditure. The problem with this choice is determining the opportunity cost of each area, as there is the problem of asymmetric information between buyers and sellers and the high cost of studies to determine this.

The lack of long-term planning in cases of PES paid for by the government is a problem, as it would result in the program ending whenever government policy changed, not continuing a successful and beneficial idea for society. However, when it comes to private financing, this is not a concern, because as long as the service is provided, users will be willing to pay.

For Peru (2010), in order for a PES to be successful, effective and sustainable, dedicated and permanent work is needed to create spaces for inter-institutional participation, incorporating civil society and strengthening institutions, structures and organizations, so that the implementation of the PES mechanism is effective and sustainable. Awareness-raising, communication and environmental education need to be worked on, as well as capacity-building on related issues.

There are several aspects that make PES programs efficient and effective. Among them: additionality, non-transfer of degrading activities, permanence, reasonable transaction costs, flexibility of payments.

6.2 Social aspects

The correlation between PES schemes and equity is limited and needs special attention, since the main focus of the instrument is environmental conservation. In order to give the instrument a dimension of justice and equity and poverty reduction, the institutional design of the program needs to be thought out with this in mind. Often, for the social side to be taken into account, the environmental side has to be sacrificed a little, i.e. the level of conservation is reduced a little so that certain actors can be included in the program.

According to Antunes (1999), the preservation and sustainability of the use of environmental resources must be seen in a way that ensures the quality of life of human beings who undoubtedly need to use the various environmental resources to guarantee their own lives, but always in a rational way since they are limited.

Changing behavior gives environmental protection the same level of importance as other social and economic values protected by the legal system. In order for this concept to be understood and applied, we must change the way we see nature, since it has its value independent of its usefulness to human beings.

According to Capra (1996), human beings are just one particular thread in the web of life, in which there is a network of phenomena that are interconnected and interdependent.

When drawing up socio-environmental policy, policymakers should consider that even those who are not the focus of the program should not be further harmed by the program's existence.

According to IPEA (2010), PES programs can lead to a win-win situation in cases where the neediest people live in areas of environmental interest, transferring resources to the poorest through payments, as well as stimulating organization and the development of more sustainable work practices. However, the opposite may happen, with the most vulnerable groups becoming even more marginalized. This can happen, for example, in cases where small rural producers do not have title to the land, or do not have the capacity to guarantee the perpetuity of the environmental service.

Pagiola, Arcenas and Plantais (2005) came to the conclusion that transition costs represent much greater obstacles to the participation of poorer communities than the limitations of the community itself. This is because it may be that in order to include poor communities that are not in areas of great interest to the program, but which would represent a social improvement for the region, those responsible for the program would spend more on technical advice and environmental cost/benefit. No major statistical differences were found between the participation levels of financially better-off and poorer communities.

For Engel, Pagiola and Wunder (2008), the basic assumption of the PSA is that none of the parties involved is harmed. Either there is an improvement or there is no agreement, even if this improvement means that only one party benefits while the other remains at the same level as before. The big question is how beneficial this agreement can be for poverty alleviation.

Another significant point that could be a consequence of the development of PES schemes in the country is the growing participation of representatives of rural producers in environmental discussions, in which official environmental bodies and NGOs have traditionally played a central role. As potential win-win situations are created, a more favorable environment is created for these discussions.

For Eloy, Coudel and Toni (2013), as much as PES programs encourage land use based on agricultural production and local knowledge, this will only have an effect for small landowners if they have access to the technical-scientific networks that structure local institutional arrangements. In other words, they need to make a planning and political coordination effort that is often beyond their capacity.

Involvement with family members, surrounding communities and local authorities, both in relation to the "core business" of PES projects and in relation to the dissemination of information about them, is easier, the more central the role of the certifying bodies becomes in terms of guaranteeing the logic of sustainable development for these projects, seeking the incorporation and deployment of the project's benefits at local level, through more dynamic and transparent social participation processes. In relation to these two points, the major challenge for the actors involved in development is to ensure that the project's benefits are incorporated into the local context.

Eloy, Coudel and Toni (2013) say that PES schemes can underutilize local agricultural systems if the focus is exclusively on conserving or restoring native forests. This is undesirable from the point of view of equity and fairness, since small farmers may not be able to adjust to the standards set by the program.

7 CONCLUSION

As seen throughout this paper, payment for environmental services has great potential to influence people's behavior and, consequently, is an interesting instrument for stimulating the conservation and/or restoration of riparian forests. In this sense, although it is not the only possible mechanism to be used, a payment for environmental services program is very suitable for this purpose.

PES seeks to soften the impacts caused by humans on the environment. If the legislation is being respected, PES can be seen as an award, a reward for the desired behavior. It is a conservation instrument that is increasingly being used due to the good results already achieved in pioneering cases. It can also be a social program, in the sense of increasing family income. The great contribution of PES is the transformation of practices that are disadvantageous in the private sphere, but socially desirable, into individual advantages, inducing their adoption and therefore improving environmental conservation.

We must not forget that PES programs are very important for encouraging the conservation and restoration of riparian forests, but it is unlikely that a producer will stop producing in order to live off the sale of environmental services. Not only would this not be part of their range of knowledge, but it would also leave them in a vulnerable situation insofar as they depend on the existence of a buyer of the service for their survival. On the other hand, it's also hard to imagine payment for environmental services lasting indefinitely. Therefore, what PES programs should encourage is a change in the production pattern of the target landowners, so that in the medium term they can make a living from their more sustainable production, without depending on external contributions from PES programs.

From the study presented here, it can therefore be concluded that PES programs are extremely important for threatened areas of high ecological value and that the mechanism can be interesting in that it has the potential to reveal and provide greater cost-effectiveness to the program, but this will depend on local socio-economic conditions and its design, in view of all that has been said here, that payment for environmental services programs are considered precursors of alliances between the public and private sectors with a common goal: environmental conservation, which is essential for current and future economic development.

Finally, it is worth mentioning that the mechanism of payments for environmental services has the potential to improve its cost-effectiveness as it becomes adopted as a permanent tool for environmental conservation and restoration, and is carried out frequently. Over time, both the financier of the PES program and the landowners involved become better at dealing with the process, using the lessons learned to optimize the results of the program. It is therefore hoped that the analysis of the scientific literature on the subject of PES presented in this monograph will indicate elements to be improved in order to improve efficiency and equity in payment for environmental services projects.

REFERENCES

ALMEIDA, L.T. **Politica Ambiental: uma analise econômica**. Campinas, SP: Sao Paulo: Fundagao Editora da Unesp, 1998.

NATIONAL WATER AGENCY. **Operational Manual for the Water Producer Program.** 2ª Edition. Brasilia: ANA, 2012

ANTUNES, P.B. **Direito Ambiental**. 3. ed. Rio de Janeiro: Lumem Juris, 1999.

BECK, U. **Risk society.** Towards another modernity. Trad. Sebastiao Nascimento. 1. ed. Sao Paulo: Editora 34, 2010.

BOURDIEU, P. **O poder simbolico**. 7. ed. Rio de Janeiro: Bertrand Brasil, 2004.

BORNER, J.; HOHNWALD, M.; VOSTI, S. Critical Analysis of Options to Manage Ecosystems Services in the Andes/Amazon Region. In: **A Situation Analysis to Identify Challenges to Sustainable Management of Ecosystems to Maximize Poverty Alleviation:** Securing Biostability in the Amazon/Andes (ESPA-AA). 2008

BRAZIL. Institute for Applied Economic Research. **Research on payment for urban environmental services for solid waste management.** Brasilia: IPEA, 2010

BRAZIL. **Constitution of the Federative Republic of Brazil.** Brasilia, 1988. Available at: http://www.planalto.gov.br/ccivil_03/constituicao/constituicao.htm. Accessed on: January 03, 2015

CAPRA, F. **The Web of Life:** A New Scientific Understanding of Living Systems. 11 ed. SP: Editora Cultrix, 1996

COASE, R. The Problem of Social Cost. **Journal of Law and Economics**, n.3, p. 1-44, october, 1960

COSTA, R.C. Payments for environmental services: limits and opportunities for the sustainable development of family farming in the Brazilian Amazon. **Thesis (PhD) -** Postgraduate Program in Environmental Science, University of Sao Paulo, Sao Paulo, 2008.

ELOY, L.; COUDEL, E.; TONI, F. Implementing Payments for Environmental Services in Brazil: paths for critical reflection. **Sustainability in Debate** - Brasilia, 4 (1): 21-42, 2013.

ENGEL, S.; PAGIOLA, S.; WUNDER, S. Designing payments for environmental services in theory and practice:An overview of the issues.
Ecological Economics, 65: 663-674, 2008.

FOOD AND AGRICULTURE ORGANIZATION. **Payment Schemes for Environmental Services in Watersheds.** Land and Water Discussion. Paper 3. Rome, 2004.

FOLETO, E. M.; LEITE, M. B. Perspectives on Payment for Environmental Services and Case Examples in Brazil. **Revista de estudos ambientais**, 13(1): 6-17, 2011.

FOREST TRENDS AND KATOOMBA GROUP, 2008. **Payments for Environmental Services**: A Handbook. On how to get started. Available at: http://www.katoombagroup.   org/ documents/events/event33/Payments_for_Environmental_Services.pdf.   Accessed: January 2, 2015.

GUEDES, F.B.; SEEHUSEN, S.E. **Payments for Environmental Services in the Atlantic Forest**: lessons learned and challenges. Brasilia: MMA, 2011

GUSMAO, A. V. P. 2008. **Global environmental problems and compensation for environmental services as an alternative for protection of social capital and ecological.**Available   at: <http://jus2.uol.com.br/doutrina/texto.asp?id=6341>. Accessed: January 2, 2015.

HERCOWITZ, M. E; WHATELY, M. Serviços ambientais: conhecer, valorizar e cuidar: subsidios para a protegao dos mananciais de Sao Paulo. **Instituto Socioambiental**, 2008.

HEYMAN, J., ARIELY, D. Effort for payment. A tale of two markets.**PsychologicalScience15**:787-793, 2004.

IPAM. **Amazon Environmental Research Institute**. 2013. Available at: http://www.ipam.org.br/saiba-mais/O-que-e-e-como-surgiu-o-REDD-
/3 Accessed on: January 2, 2015.

IPEA - INSTITUTE FOR APPLIED ECONOMIC RESEARCH. **Environmental sustainability in Brazil**: biodiversity, economy and human well-being. Brasilia, DF: Ipea, 7, 2010.

KAWAICHI, V.M. An analysis of countries' environmental public policies and the adoption of Payment for Environmental Services in Brazil. **Graduation Monograph.** University of Sao Paulo, 2009.

KOSOY et al. Payments for Environmental Services in Watersheds: Insights from a comparative study of three cases in Central America. **Ecological Economics.** 61(2-3): 446-455, 2007.

LANDELL-MILLS, N.; PORRAS, T.I. Silver bullet or fools' gold? A global review of markets for forest environmental services and their impact on the poor. Instruments for sustainable private sector forestry series. **International Institute for Environment and Development,** London, 2002.

LEONARDO, H.C.L. Soil and water quality indicators for evaluating the sustainable use of the Passo river basin, western region of the state of Paraná. **Dissertation (master's degree in forestry sciences).** 90f. Luiz de Queiroz School of Agriculture-ESALQ-USP. Piracicaba. 2003.

LOUREIRO, W. ICMS Ecologico, uma experiência brasileira de pagamentos por serviços ambientais. Belo Horizonte: Conservagao International - Sao Paulo: Fundagao SOS Mata Atlantica - Curitiba: **The Nature Conservancy (TNC),** 2008.

MAY, P.H.; GELUDA, L. Payments for ecosystem services to maintain sustainable agricultural practices in micro-watersheds in the northeast and northwest of Rio de Janeiro. **Congress of the Brazilian Society of Ecological Economics,** Brasilia - DF, Brazil. 2005.

MOTA, J.A. **O valor da natureza**: Economia e politica dos recursos ambientais.Rio de Janeiro. Ed. Garamond. 2006.

MURADIAN, R.; CORBERA, E.; PASCUAL, U.; KOSOY, N.; MAY, P.H. Reconciling theory and practice: An alternative conceptual framework for understanding payments for environmental services. **Ecological Economics,** 69: 1202-1208, 2010.

NUSDEO, A.M.O. **Payment for environmental services:** sustainability and legal discipline. Sao Paulo: ed. Atlas, 2012

ORTIZ, R.A. Environmental economic valuation. In: MAY, P. H.; LUSTOSA, M.C.; VINHA, V. (org). **Environmental economics**. Rio de Janeiro. Ed. Elsevier, 2003.

PAGIOLA, S.; ARCENAS, A.; PLATAIS, G. Can Payments for Environmental Services Help Reduce Poverty? An Exploration of the Issues and the Evidence to Date from Latin America. **World Development:** 33(2): 237-253, 2005.

PERU, M. **Compensation for ecosystem services**: Lessons learned from a demonstrative experience. Lasmicrocuencas
Mishiquiyacu, Rumiacu y Almendra de San Martin, Peru. Lima: Ministry of the Environment, 2010.

REZEK, F. **Direito Internacional Publico:** curso elementar. Sao Paulo: Saraiva, 2008.

ROJAS, M.; AYLWARD, B. What are we learning from the experience with environmental services markets in Costa Rica? A review and critique of the literature. **Markets for Environmental Services.** London: IIED, n°. 19, 112p., Aug2005. Available at: <
http://www.iied.org/pubs/pdfs/12534IIED.pdf>. Accessed on: November 30, 2014

ROMEIRO, A. R. **Economia ou economia política da sustentabilidade.**Unicamp, Campinas, 102, 2001.

SANTILLI, M.; MOUTINHO, P.; SCHWARTZMAN, S.; NEPSTAD, D.; CURRAN, L.; NOBRE, C. **Tropical deforestation and the Kyoto Protocol:** an editorial essay. Amazon Environmental Research Institute. 2000.

SANTOS, A.C.; GUIMARAES, R. M. A. State, democracy and public policies. IN SANTOS, Antonio Carlos dos (Org.). **Philosophy & Nature:** debates, clashes & connections. 2ª Ed. Sao Cristovao - SE: Editora UFS, 2010.

SEEHUSEN, S.E. Can payment for ecosystems services contribute to sustainable development in the Brazilian Amazon? With case study from the Rio Capim Pole of Proambiente. **Master's thesis.** Freiburg, 2007.

SEROA DA MOTTA, R. **Manual for the Economic Valuation of Environmental Resources.** IPEA/MMA/PNUD/CNPq. Rio de Janeiro, September 1998.

SOMMERVILLE, M. M.; JONES, J, P. G.; MILNERGULLAND, E.J. A revised conceptual framework for payments for environmental services. **Ecology and Society**, 14(2): 34, 2009.

SPIRONELLO, R.L. Anthropic-environmental zoning of the municipality of Ipora do Oeste - SC: a contribution to reflection and decision-making in the context of hydrographic micro-basins. University of Sao Paulo, Dept. of Geography,

Postgraduate Program in Geography, **Thesis (Doctorate)**, Sao Paulo, 2008. 161 p.

TRES, D.R. **Guide to payments for environmental services for watershed protection.** Associação Terceira Via, Joanopolis/SP, 2011.

VASCONCELLOS, C.S. **Avaliagao concepgao dialetica** - libertadora do processo de avaliagao escolar. Sao Paulo: Libertad, 2005.

VARELA, C.A. **Instrumentos de Políticas Ambientais Cases of Application and their Impacts.** EAESP/FGV/NPP - Nucleo de Pesquisa e Publicagoes. Research Report no. 62, 2001.

VEIGA NETO, F.C. The construction of environmental services markets and their implications for sustainable development in Brazil. **Thesis (Doctorate).** Postgraduate Program in Agriculture and Society Development - CPDA/UFRRJ. Rio de Janeiro, 2008

WUNDER, S. The Efficiency of Payments for Environmental Services in Tropical Conservation. **Conservation Biology.** v. 21, p. 48-58. Payment for environmental services: some nuts and bolts, 2007

WUNDER, Sven; ENGEL, Stefanie; PAGIOLA, Stefano. Taking stock: a comparative analysis of payments for environmental services programs in developed and developing countries. **Ecological economics,** 65, 834-852, 2008.

WUNDER, S.; WERTZ-KANOUNNIKOFF, S. Payment for Ecosystems Services: A New Way of Conserving Biodiversity in Forests. **Journal of Sustainable Forestry,** 2009.

WCED. **Our Common Future.** 1987. Available at:http://www.un-documents.net/wced-ocf.htmAccessed November 30, 2014

Printed by Books on Demand GmbH, Norderstedt / Germany